Lapwing
Vanellus vanellus
cinerea
Common Reed
Phragmites australis
Great White Egret
Egretta alba
Graceful Prinia
Prinia gracilis
Little Egret
Egretta garzetta
Zitting Cisticola
Cisticola juncidis
Black-crowned Night Heron
Nycticorax nycticorax
Egyptian White Lotus
Nymphaea lotus
Clamorous Reed Warbler
Acrocephalus stentoreus
Squacco Heron
Ardeola ralloides
AF898544

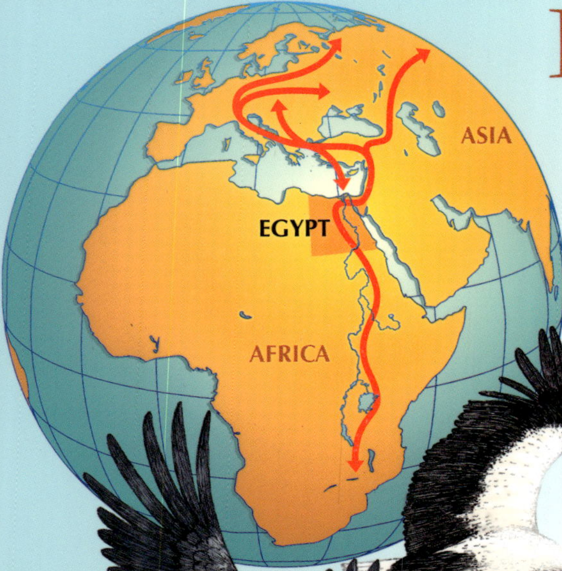

Foreigners flock to Egypt, both tourists and birds alike. Egypt is a unique land bridge creating one of the world's largest bird migration systems, a route referred to as the African-Eurasian Flyway.

In fall and again in spring, one can witness the passage of over five million birds all along the Nile on their journeys between Europe and Asia to central and southern Africa. The Nile is the life source for all of Egypt and creates a wealth of wetlands that are dense and diverse with bird species, an abundance of prey and lush vegetation. As the seasons change and one travels from the Delta to the southernmost regions of the country, different bird species appear. These migrants feed and rest among the numerous resident species that call the Nile Valley home.

White Stork
Ciconia ciconia

Eurasian Crane
Grus grus

Great White Pelican
Pelecanus onocrotalus

Of the 485 species of birds found in Egypt, over 170 are residents.

AUTUMN APPOINTMENTS

The first movement south of hundreds of thousands of birds begins in August with great crowds of impressive **White Stork**, V-shaped flocks of the **Eurasian or Common Crane**—the tallest flying bird in the Middle East—and hordes of **Great White Pelican** filling the sky. Both storks and cranes fly with their necks and legs fully extended, while the necks of pelicans, like those of herons, are distinctively tucked into a curve. Like many other birds, all three of these species are moving from their summer breeding grounds in Europe to winter feeding areas in Africa. They soar and glide on warm air thermals, down through the Levant (the eastern coast of the Mediterraenan Sea) and Sinai into the Nile Valley and then onward to the south.

Akhet was the ancient Egyptian name for the season of inundation, from August to September, when historically the Nile waters rose, flooding and fertilizing the farmlands with a rich silt. With the construction of various dams and barrages in the 19th century and the Aswan High Dam in 1970, the river was tamed and no longer floods. Yet the level of the Nile still fluctuates and nurtures numerous and vital wetlands where a gorgeous assortment of waterbirds can thrive. Twelve species of heron occur regularly in Egypt including the **Gray Heron**, **Great White Egret**, and **Little Bittern**, all from the family *Ardeidae*. Herons have good eyesight, long legs, necks which can kink into an S-shape, and long spear-like bills, which together make them formidable hunters of fish, amphibians, and other aquatic creatures. On a walk along the Corniche in Luxor or a boat ride in a felucca at Aswan, one is sure to spot a heron standing boldly beside the water patiently seeking its next meal.

Some resident birds like the **Spur-winged Plover** are noisy and conspicuous while others, like the **Squacco Heron**, the **Clamorous Reed Warbler** or the **Purple Gallinule**, will try to conceal themselves in reed-beds.

Spur-winged Plover
Vanellus spinosus

WINTER APPEARANCES

By winter many birds will have lost their distinctive summer plumages and molted into much duller colors. For example, the male **Nile Valley Sunbird** swaps its glossy blue-green coat and brilliant golden belly for pale grey and yellow. Nevertheless, there are always other delights to be found.

With an abundance of fish in the river and fresh water lakes, a rare and rewarding sighting in northern Egypt is the **White-breasted Kingfisher** with its array of colors and disproportionately large beak, while the black and white **Pied Kingfisher** occurs all along the Valley catching **Nile Tilapia** fry and dragonflies.

A winter arrival to brackish lagoons, mainly in the north, is the **Greater Flamingo**. Flocks of hundreds settle into wetland areas such as Lake Qarun in the Fayoum region, a Natural Protectorate. A pinky-white species, the Flamingo has a distinctive, massive beak for just one purpose: filter feeding. With the aid of a large tongue and hair-like lamellae, water and soft mud are pressed and strained to leave a rich, edible mixture of algae, mollusks and crustaceans. The **Eurasian Spoonbill** uses its spatula-shaped beak to sweep shallow waters for tiny fish and crustaceans while the **Glossy Ibis** probes in mud for its prey with a long, decurved bill. The **Pied Avocet's** up-curved, delicately shaped beak adds to its lovely appearance. Although found in the same habitats, none of these species compete for food as their differently shaped bills necessitate differing feeding behaviors. Similarly, the **Black-winged Stilt** has a longish, needle-fine, straight beak, which, coupled with its extraordinarily long legs, enables it to wade and feed in deeper water.

Osprey enjoy the Nile's abundance of fish, migrating west in the winter from the Red Sea to the Nile waters of Aswan. Aswan and Lake Nasser are opportune places to see other residents, such as **Yellow-billed Kite** and **Great Cormorant.**

COVER SPECIES:
Egyptian Goose Alopochen aegyptiacus
Eurasian Hoopoe Upupa epops
Glossy Ibis Plegadis falcinellus
Grasshopper Poekilocerus bufonius
Nile Valley Sunbird Anthreptes metallicus
Pied Kingfisher Ceryle rudis
Lesser Kestrel Falco naumanni
Lanner Falcon Falco biarmicus
Painted Lady Vanessa cardui
Little Green Bee-eater Merops orientalis
Common Bulbul Pycnonotus barbatus
Philae Temple
Kentish Plover Charadrius alexandrinus
Senegal Coucal Centropus senegalensis
Nile Tilapia Oreochromis niloticus
Purple Heron Ardea purpurea
Common Moorhen Gallinula chloropus
Eurasian Coot Fulica atra
Masked Shrike Lanius nubicus
Pintail Ducks Anas acuta
Spotted Crake Porzana porzana
Giant Milkweed Calotropis procera
Purple Gallinule Porphyrio porphyrio
Egyptian Plover Pluvianus aegyptius
Painted Snipe Rostratula benghalensis
Senegal Thick-knee Burhinus senegalensis
Laughing Dove Streptopelia senegalensis
Little Bittern Ixobrychus minutus

BIRDS OF THE NILE VALLEY

BIRDS OF THE NILE VALLEY

AN AUC PRESS NATURE FOLDOUT

Yellow-billed Kite
Milvus aegyptius

Black-winged Kite
Elanus caeruleus

Yellow Wagtail
Motacilla flava

Crested Lark
Galerida cristata

Egyptian Vulture
Neophron percnopterus

Barn Swallow
Hirundo rustica

Pallid Swift
Apus pallidus

Barbary Falcon
Falco pelegrinoides

FORAGING AHEAD

As the floods receded in ancient times, Egypt entered its growing season, ***Peret***, which ran from October through to March. This was the time to sow and plant the fields. Today, with reliable irrigation canals, Egyptian farmers can have up to three crops in a year. Flooded farmlands are ideal for bird watching. **Yellow Wagtail** and **Crested Lark** are easily identifiable species throughout the Valley and no field is complete without flocks of **Cattle Egret** seeking a meal in the freshly plowed soil. Another resident, the **Black-winged Kite**—a bird of prey with beautiful, piercing, ruby-red eyes—may also be hovering overhead searching below for small rodents, lizards, and insects.

Cattle Egret, *Bubulcus ibis*, in a field on the west bank of Luxor.

SPRING'S DELIVERANCE

With springtime, March and April were the months in which the ancient Egyptians looked forward to harvesting abundant crops if the river and weather gods had been kind. Chirps and chirrups fill the air as nests abound in every habitat, hollow, and homestead alike. The resident **Eurasian Hoopoe**, with its regal crown of fanned feathers, will make its nest in the hole of a tree or wall, while the elusive, nocturnal **Senegal Thick-knee** may select a roof-top nest even in the midst of Cairo. **Pallid Swifts** will conduct their aerial courtships while **Little Green Bee-eaters** may burrow out their nests in sandy banks, as do Kingfishers. Aswan's Botanical Garden on Kitchener Island and Cairo's Giza Zoo have delightful colonies of white **Little Egrets** nesting in Nile Acacia trees, gargling noisily with their distinctive croaks. They can be distinguished from other herons by their black bills, black legs and yellow feet. Among city dwellers, like the cooing **Laughing Dove** (or **Palm Dove),** is the grey and black **Hooded Crow**, a common breeding resident throughout the Valley, which often nests in the crowns of Date Palms.

Spring sees migrating birds, such as the **Egyptian Vulture**, passing through the Valley once again as they leave their winter quarters deep in the African continent. **Masked Shrikes,** known as "butcher birds" for impaling their prey on thorns, are easily recognizable as they migrate through. Butterflies, such as the **Painted Lady,** are also heading northwards and are especially abundant in open countryside.

The melodic songs of the **Graceful Prinia** and **Common Bulbul** fill the air while the Egyptian red-bellied race of **Barn Swallow**, so common as to go almost unnoticed, reminds one of its presence every dawn and dusk by swarming with a lovely hum in trees along the Nile.

Hooded Crow
Corvus cornix pallescens

SUMMER EVENINGS

With rising temperatures comes the summer drought or *Shemu* as it was known to the ancients. People, animals, and birds become creatures of the night. As the sun sets, a few charming residents, including **Painted Snipe** and **Black-crowned Night Heon**, may appear along the water's edge. The **Egyptian Nightjar** spends its days lying invisibly on desert sands but takes to the air at dusk with a moth-like flight. In the night sky, one would be very lucky to see the magnificent "eared" **Pharaoh Eagle Owl**, as it prefers desert regions, but some have been sighted nesting in the vicinity of the Giza and Saqqara Pyramids. The whiter **Barn Owl** can be easily distinguished passing silently overhead while a fleeting flash in the air above may not be a bird at all but rather the only mammal capable of bird-like flight: the bat. There are over 20 bat species in Egypt, the largest being the **Egyptian Fruit Bat**.

Owls are one of the many bird species often depicted in hieroglyphs. Above, the **Barn Owl** as a glyph, followed by the plume of an **Ostrich**, a falcon, a **Pintail Duck**, a **Glossy Ibis**, and a **Lappet-faced Vulture.**

House Sparrows, *Passer domesticus,* may nest in temple walls among the hieroglyphs like this pair at Dendera's Hathor Temple.

A **Brown-necked Raven,** *Corvus ruficollis,* overlooks a sculpted relief of the raptor god Ra-Horakty.

ANCIENT NATIVES

Wander into nearly any temple or tomb whose walls are graced with paintings and hieroglyphs and you will find yourself a keen bird watcher. The ancient Egyptians depicted over 100 different bird species in their art and as symbols in their writing. And in the open courtyards of antiquities such as Luxor Temple and Medinet Habu, look up and you will often find resident **House Sparrows** nesting or a **Brown-necked Raven** gazing down from atop the lofty temple walls.

Birds in ancient Egypt played a key role in everyday life. Some species were hunted while others were tenderly mummified like the pharaohs. Others inspired hieroglyphic symbols or were deified and venerated in wall paintings and sculpted reliefs.

While some species like the **Ostrich, Lappet-faced Vulture,** or **Sacred Ibis** (which embodied the God of Wisdom, Thoth) are reverently depicted in ancient art, such birds have sadly become rare or locally extinct in Egypt due to overhunting and loss of habitat. Pollution has transformed the Nile waters, where Papyrus and Lotus swamps have long disappeared, forcing some migrants to adapt or seek sanctuary elsewhere. Today, the sacred artwork of the past is a reminder of the natural splendors of Egypt that once were.

IMPORTANT BIRD AREAS OF EGYPT

Over the last century, Egypt's internationally important wetlands have suffered drastically from habitat destruction, pollution, drainage, overfishing, and overhunting. Egypt's Natural Protectorates and Important Bird Areas are essential to conservation of vital habitats and Egypt's biodiversity. The Important Bird Areas Program (IBA) identifies and conserves areas of global significance to birds and biodiversity.

1 Lake Bardawil
2 Zaranik
3 El Mallaha
4 Bitter Lakes
5 Lake Manzala
6 Lake Burullus
7 Lake Idku
8 Lake Maryut
9 Lake Qarun
10 Wadi El Rayan
11 Wadi El Natrun
12 Upper Nile
13 Aswan Reservoir
14 Lake Nasser
15 Hurghada Archipelago
16 Tiran Island
17 Wadi Gimal Island
18 Qulan Island
19 Zabargad Island
20 Siyal Islands
21 Rawabel Islands
22 Nabaq
23 Gebel Elba
24 Abraq
25 St. Katherine
26 Gebel Maghara
27 Quseima
28 Wadi Gerafi
29 El Qasr Desert
30 Suez
31 Gebel El Zeit
32 El Qa Plain
33 Ras Muhammad
34 Ain Sukhna

The American University in Cairo Press
www.aucpress.com

ISBN 978-977-416-579-5

Scientific Consultant: **John Wyatt**
Author / Illustrator: **Dominique Navarro**
www.dominiquenavarro.com
Design: **Dawn N. Ericson**
www.mantapublications.com
Copyright © 2012 Dominique Navarro

All rights reserved.
Dar El Kutub No. 11836/12
ISBN 978-977-416-579-5
Printed in Egypt